Volume 2 of Moonwalkers™ Series

STAR TRUCK

Untold History of the Space Shuttle

By

John Getter

www.JohnGetter.com

Star Truck

John Getter

MOONWALKERS™

Volume 2 – STAR TRUCK
Untold History of the Space Shuttle

Electronic Edition First Published April 2011
Electronic ISBN 978-1-4524-4703-2

ISBN-13: 978-1463624156
ISBN-10: 1463624158

Premiere Projects
eBook Publishing & Promotions

Premiere Projects
Henderson, NV 89002
24/7 GoogleVoice: 702-900-2176
www.PremiereProjects.com

Moonwalkers™ Series

By John Getter

Amazon Best Sellers

Available in Paperback and E-Books

More Information at www.JohnGetter.com

Volume 1 – TO THE MOON

Untold Stories of the Space Race

These are the untold stories of the space race and who would put the first man on the moon from insiders in the Russian and American space programs. Read the jaw-dropping facts as shared by Emmy-winning and longtime space exploration reporter John Getter. These are the facts and secrets previously known only to a handful of technicians, cosmonauts and astronauts.

John shares why the then-Soviets beat the Americans everywhere but to the moon. Get answers to questions such as: Why did Yuri Gagarin and the early Russian cosmonauts never land in their capsules? Who were the Mercury 13? And what does mowing the lawn have to do with winning the space race?

This book reveals why the American "better" put the first and only men to walk on the moon while the Russian "good enough" led to so many firsts. Enjoy color photographs from John's personal collection as well as NASA and Soviet archives. Join John as he shows how the space race changed – and continues to change – the human experience.

Volume 2 – STAR TRUCK

Untold History of the Space Shuttle

In Volume 2 of his Moonwalkers™ series, author John Getter once again uses his experience as NASA's reporter-of-choice for pool coverage, private pilot and personal friend to many of the astronauts to reveal the little-known stories behind the history of the space shuttle.

Join John as he takes the reader on a journey through the joys and frustrations of the human adventure of the space shuttle and exploration, from its inception during the first American's walk on the moon to the last shuttle flights. Share the humor of the "cola wars" in space and the "toilet tours." Learn more about the shocking revelations and cover-ups that followed the Challenger and Columbia disasters. Experience the emotions as John describes the accidents and what the investigations revealed really happened to the crew as their shuttles were destroyed. And examine the dangerous and potentially devastating effects resulting from the decision to end the space shuttle program prematurely.

This e-book includes 30 color photographs from John's personal collection as well as NASA archives. "Star Truck" is a great follow-up to John's best-seller Moonwalkers™ Volume 1 "To The Moon: Untold Stories of the Space Race."

Volume 3 – TIMELINE

Untold Stories of Space Exploration

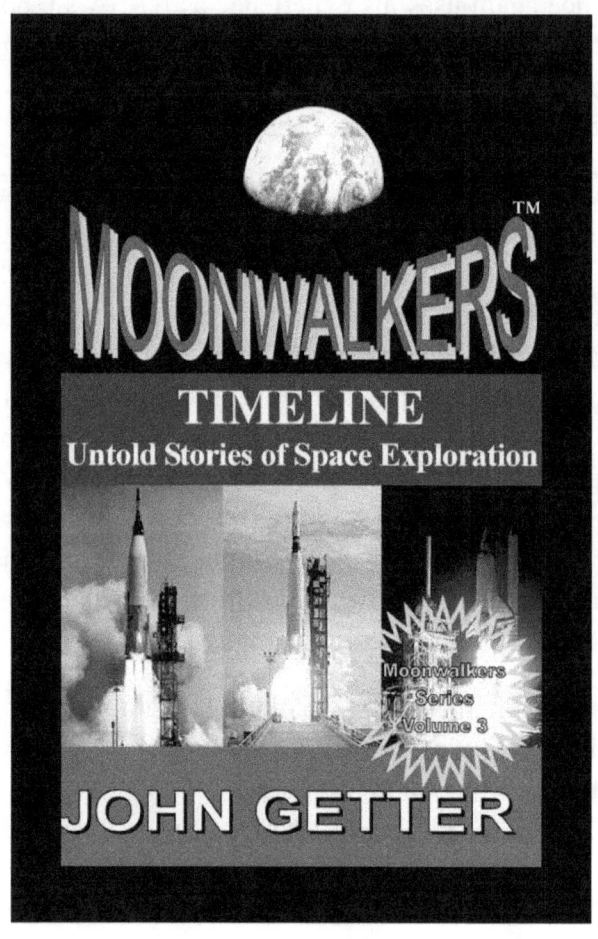

In Volume 3 of his Moonwalkers™ series, author John Getter once again uses his experience as NASA's reporter-of-choice for pool coverage, private pilot and personal friend to many of the astronauts to reveal the little-known stories behind the history of space exploration.

Join John as he takes the reader on a day-by-day journey through American and Soviet space exploration from the birth of NASA in 1958 and the first man in space – a Soviet – through the 1972 last walk on the moon – an American. Discover why Russian cosmonauts urinate on a vehicle tire before every flight and what became the politically incorrect catchphrase all astronauts hear as they prepare to climb into their spacecraft. Learn more about the triumphs, disasters and cover-ups that were such an integral part of the Cold War race to the moon.

This e-book includes 37 color photographs from John's personal collection as well as NASA and Soviet archives. A great follow-up to John's best-sellers Moonwalkers™ Volume 1 "To The Moon: Untold Stories of the Space Race" and Moonwalkers™ Volume 2 "Star Truck: Untold History of the Space Shuttle."

Star Truck

A portion of the net proceeds from this e-book will be donated to the Challenger Center for Space Science Education. This not-for-profit education organization was created in 1986 by the families of the astronauts from tragic Challenger Space Shuttle mission STS-51-L. This nonprofit is dedicated to the educational spirit of that mission and develops Challenger Learning Centers and other educational programs worldwide to continue the mission of engaging students in science and mathematics education. To learn more go to:

www.Challenger.org

Table of Contents

Photographs

ABOUT THE AUTHOR

John Getter earned his reputation as a pre-eminent storyteller early in his broadcasting career. While in Cincinnati, his profiles of interesting and original people connected him with a variety of subjects. Interviews ranged from Lou Jacobs, who for decades was the image of the Ringling Brothers, Barnum and Bailey Circus, to physicist Edward Teller, who admitted he hated being called the father of the hydrogen bomb, to Neil Armstrong, first human to walk on the moon.

Building on that remarkable start, John was enthralled when he saw the early tests of what would become the space shuttle. He moved to Houston and spent the next several years covering the manned space program.

John is also a private pilot, so he combined his knowledge of aviation with the love of a good story and soon became known as one of the most respected reporters covering exploration. He flew the space shuttle simulator, worked in real spacesuits, participated in scientific tests that later flew in space and spent more than 200 minutes in zero gravity.

He ultimately became NASA's reporter-of-choice for pool coverage of activities aboard the famous "Vomit Comet" or Zero-G airplane. He flew with countless astronauts, mission specialists and politicians. He and his videographer helped several astronauts learn how to shoot video in weightlessness before they ventured to space.

John Getter Weightless in Vomit Comet

The crew of the final ill-fated flight of the Challenger considered John a personal friend. He built on these connections to become personally acquainted with every surviving moonwalker and was a major player hosting each of them in the official 20th anniversary celebrations of the Apollo program at Johnson Space Center.

His reports were regularly featured by CBS News and CNN and seen in more than 135 countries. His reporting included coverage from all across America, Central and South America, Europe and Russia. In addition, he covered news based in Washington, D.C., that included the White House, presidential transitions and Congress.

Recognizing his experience and insights, John was recruited to join SPACEHAB as a senior vice president for Space Media Inc. He developed and managed plans for construction of what would have been the first dedicated media facility aboard the International Space Station. With staff in the United States and Russia, he was involved in several commercial space missions using the shuttle, Soyuz, MIR and International Space Stations.

John has also consulted with major corporations, industry and political leaders on effective communications skills, media understanding and, of course, storytelling.

He now lives in Las Vegas where he enjoys the desert, is very active in rescuing Doberman Pinschers, produces news and television content, consults with clients and still tells stories such as these.

To contact John for keynotes, inspirational talks, effective communications and consultations:

John Getter

6380 Golden Goose Lane

Las Vegas, NV 89118

702-531-7486

John@JohnGetter.com

www.JohnGetter.com

Star Truck

A Magic Moment

More than a billion people were glued to their televisions in July 1969 amazed to see the fuzzy and stark images.

Two Americans named Neil Armstrong and Buzz Aldrin were walking on the moon.

The great news anchor Walter Cronkite was reduced to tears by the sight. Stores were empty and streets were quiet in countless countries. It was a truly unique moment in human history that was recognized as such by most everyone in what would be termed the civilized world, and they wanted to watch it on live television.

It was not just that two men were walking on solid ground not a part of Earth.

It was not just that we could now gaze at the moon and wonder exactly where they stood.

It was amazing, compelling and irrefutable proof the United States had won the space race.

The goal of landing "a man on the moon and returning him safely to the Earth in this decade ..." had animated the Cold War fight for the hearts and minds of the world, especially the Third World. The United States and the Soviet Union each believed winning the race would make it obvious which political system was superior because the winner of this race would have the best technology to build the future.

This was really big stuff.

Aldrin Becomes Second Man to Walk on Moon

So big in fact, that Soviet leaders nearly immediately canceled their own lunar program and focused instead on dominating what is called low Earth orbit, command of space between 100 and 1,240 miles above the ground. Unknown to most everyone, the United States, too, almost canceled the rest of the Apollo missions to do the same thing.

President Nixon would cancel three of those planned lunar flights, Apollo 18, 19 and 20. He would approve plans to try for the first American space station. The Skylab used leftover rocket boosters and Apollo capsules to counter the growing presence of Soviet space stations. Nixon also approved the first joint mission with the Soviets. It was called Apollo-Soyuz in America and Soyuz-Apollo in the East. The Soyuz capsule had been developed as part of the failed Soviet lunar program.

Even as Armstrong and Aldrin made history, quietly and in unofficial secrecy, NASA and military planners would take breaks from their main task to watch the Apollo landings. Starting in 1968, they had been at their assignment for more than a year. That main task – to develop something they would call the space shuttle.

Space Shuttle: What a Concept

The early concepts of the space shuttle were based on building a different spacecraft that would use the same huge Saturn 5 boosters that took capsules to the moon. The biggest difference would be the ability to reuse the spaceship over and over and to land on a runway. An Apollo capsule, like the Gemini and Mercury capsules from America and the capsules flown by the Soviets, were good for only one flight. The new shuttle would be a glider for its return from space. Landing on a runway, it would carry four people to and from space stations and support the creation of a near-Earth infrastructure where humans would live and work.

For the next few years, engineers worked through various scenarios and designs. With its mission of commanding the skies, the Air Force was very interested in participating and helping to fund the program. With its mission of commanding the seas, the Navy was also very interested in the new ocean of space.

It was decided that technical considerations would mean the space shuttle would carry its own liquid-fueled engines that are used only to get to orbit. Their fuel would be carried in a huge tank that would be thrown away when it emptied at the edge of space. The tremendous weight of it all would necessitate the first-time use of solid-fuel rockets to get off the launch pad much like the solid rockets that helped airplanes take off from very short runways. And, the shuttle would launch

EVERYTHING the United States would put in space.

The shuttle as originally conceived had grown in size to be able to carry tons of cargo. It could readily handle a crew of eight. It could fly to nearly 300 miles above Earth. And, with all that ability it could boost the deteriorating orbit of the Skylab space station and extend its life for at least 10 years.

Perhaps the most critical decision was that America would have no functioning manned spacecraft between the early end of the Apollo program and the start of the space shuttle flights. NASA estimated it would be a gap of only a couple years at most.

The Engineers

One of the most famous names of the Apollo era was Chris Kraft. He spent his entire professional career working for NASA. Well, actually, he worked for NASA before it was NASA. Graduating from Virginia Tech in 1944, he went to work for NACA, the National Advisory Committee for Aeronautics.

In 1958, he joined the Space Task Group which was assigned the job of preparing for manned flights in space. NACA became NASA, the National Aeronautics and Space Administration, and Kraft became the first flight director in history.

Watching the remarkable cohesion, efficiency and capabilities of the mission control center in Houston today, it is important to understand that when Kraft started, there was no such thing. He invented it. An engineer by training, he focused his growing cadre of engineers and scientists on figuring out what kind of vehicles could fly in space and what it would take to manage those amazingly complex missions in the days before computers and digital processing.

Chris Kraft

They used slide rules and common sense, trial and error and unbending dedication and passion.

Kraft was the flight director for several missions beginning with Project Mercury and ending with the lunar landings. He said, "I didn't take any management course or training or anything like that," and yet he created one of the finest management organizations in history. As the Apollo program ended, his management style and skills were directed at the new space shuttle. It was not only extremely more complicated as a spacecraft. The missions it would fly were so complex they were mind-boggling.

Chris Kraft learned many things along the way to his job of taking on management of the Johnson Space Center while the shuttle was being created and first flown. Years later, he described his feelings when President Kennedy made his famous speech committing America to the moon landings within 10 years. "I thought he was crazy; there was no way!" But, there was a way and he had been a prime mover in creating and making it work. Maybe President Kennedy wasn't so crazy after all.

But, as this engineer led thousands of other engineers creating the space shuttle program, he saw unexpected problems that caused serious delays. One of the toughest involved the TPS, the thermal protection system or heat shield tiles. Most felt that once they were developed to protect the ship from a few thousand degrees of heat, all was on track.

Unfortunately, it took even longer to come up with glue that would prevent them from falling off, something that would destroy the shuttle. (It was a loss of heat shield protection that would later doom the first shuttle Columbia to its flaming demise over Texas.)

Even as planners were designating daring missions for the new spaceship, problems with tiles, engines, computers, rockets and myriad other technical issues were adding up. The second space shuttle mission was designated as the one to save the orbiting Skylab. But reality got in the way. Skylab fell from orbit in a flaming trail of destruction across the skies two years before even the first space shuttle mission could fly.

What did this series of experiences teach Kraft? He put it simply. "I've learned that engineers overestimate what they can do in the near-term and underestimate what they can do in the long-term." It is a lesson he learned the hard way. It is a lesson that has been at the core of the story of the space shuttle. It is profound and has been dubbed Kraft's Law.

The Early Days

The first space shuttle was formally called OV-101, OV as in orbital vehicle. The OV, or orbiter as it became known, is but one of the major parts of the space transportation system, or STS. The other major parts are the big external fuel tank and the solid rocket boosters attached to the side of that tank.

When it was built OV-101 was to be formally named the Constitution and be rolled out before the public for the first time on Constitution Day, Sept. 17, 1976. But, fans of the TV series "Star Trek" mounted a huge write-in campaign demanding the first shuttle be named Enterprise. President Gerald Ford had served in WWII aboard a tender to the USS Enterprise and said, "I'm partial to the name." With that, he overrode NASA managers and the first shuttle was indeed the Enterprise. It rolled off the assembly line into public view accompanied by the actors from the TV series.

Enterprise was involved in a series of early tests. As space shuttles would be ferried atop a modified Boeing 747, the first task was to see if the ungainly stack of airplane and spaceship could actually fly safely. Then, it was taken aloft and released from the back of the airplane to test its ability to fly.

Veteran lunar astronaut Fred Haise and pilot Gordon Fullerton were the first to fly her. Haise was the lunar module pilot for the ill- fated Apollo 13 mission and was slated to command the canceled Apollo 19 that would have included a moonwalk. He also was slated to command the second shuttle mission to save Skylab. But, when shuttle delays led to Skylab's demise, he planned to retire from the space agency. He stayed long enough to be the first man to fly a shuttle in the atmosphere.

Enterprise Free Flight

"She flies like a brick!" Haise remarked. To land a shuttle is not like landing any airplane most have ever flown. It descends at a vertical speed of 120 knots until just before touchdown! Near-lunar astronaut Joe Engle and Dick Truly were next. Engle was bumped from Apollo 17, the final lunar mission, to make room for Jack Schmitt, the only scientist to visit the lunar surface. He and Truly, who was part of the canceled military space station called the manned orbiting laboratory, would go on to make the second orbital flight of the shuttle.

Three more flights and landings and NASA was convinced it knew enough to fly back from space. Now, it just had to get there.

Although the Enterprise was originally planned to go to space, a number of design changes led to the decision to keep it grounded. It went on to help test the fit of the new shuttle launch pad, toured the world atop that 747 and now sits in the Smithsonian.

Since her flying days have apparently ended, Enterprise is expected to be supplanted by Discovery and continue its itinerant Earthbound journeys to various locations.

Columbia soon rolled off its California assembly line. It would become the first space shuttle to fly in space. It would have the word "first" describing it more than any spaceship in history.

STS-1: Historic, Scary and Daring

"Well, I was sittin' there on the moon one day ..." is how John Young can nonchalantly begin a story that only he can tell.

Young is the astronauts' astronaut. He first flew to space on a pair of Gemini missions where he became as famous for sneaking aboard a ham sandwich as for making the very complicated flights successful. He had been to the moon twice. First he was command module pilot on Apollo 10, a mission that did everything Apollo 11 did but land. Then, he commanded Apollo 16, spent days on the moon and drove the rover, or buggy.

To finish the moon story, "I looked out the window of our little lander. We'd been out moonwalkin' and were supposed to sleep and for some reason I just wasn't sleepy. Out there above the lunar horizon is the Earth. I put up my thumb and it disappears. I look down at the sides of our lander, the lowest and best bid, and its sides are bulging, it's moanin' and groanin', and I look out the window at that little Earth. Then, back to the lander and finally I just pulled down the shade and read my checklist. I figured I'd better focus on what's next or I'd get too worried about being so far from the house."

That remarkable ability to put aside circumstances and experiences that would overwhelm most mortals is exactly what NASA wanted in the commander of the first shuttle mission. John Young was in charge. He was also the chief astronaut and he chose as his pilot Bob Crippen. Crippen was a space rookie but considered one of the absolute best pilots and brightest minds at Johnson Space Center.

NASA had never before made the first flight of any vehicle with people aboard. It was simply too dangerous. But, the shuttle could not fly safely in automated mode and so there was no choice. "Somebody had to do it, might as well be me," joked Crippen. There were many delays and the first launch was delayed for more than a year. Significant problems with the heat shield tiles were the most common news stories for months before liftoff.

The first four flights of the shuttle were considered experimental. The ship was equipped with ejection seats and the two crew members wore the same pressure suits that protected high-altitude spy plane pilots. On STS-1, they were almost put to use.

The media covering the launch came from all around the world. We were jammed into several buildings and NASA even had to close its Johnson Space Center museum to house us. We all had a sense this was history in the making. Fed a steady diet of successes, it seemed the flight was nearly perfect. Nearly.

For launch, NASA had public affairs officers stationed at possible emergency landing sights around the world. We learned later they also had them at the hospital burn unit near Kennedy Space Center just in case the shuttle exploded or crashed right after liftoff. No one knew exactly if and how it would fly until it flew. After a two-day delay, the launch happened at exactly 7 a.m. EST, on the 20th anniversary of man's first spaceflight by Yuri Gagarin.

A launch is spectacular.

As Bob Crippen said, "That thing's just alive. They light the main engines at seven seconds before liftoff and the whole ship bends three feet under the pressure. Then, when those solids ignite (at liftoff) you know you're certainly going somewhere."

Because of its size, the more than 100-ton spaceship appears to be moving slowly at first. But, by the time it has cleared the launch tower it is already zooming upward in excess of 100 miles per hour.

For two minutes Columbia climbed. Then, in a burst of flames, the solid rocket boosters were blasted away. None of us had ever seen this, of course, and it basically scared the hell out of nearly everyone watching.

The main engines worked just fine and nine minutes later the shuttle was in space, moving at 5 miles per second and 156 miles up, and we could finally exhale.

Then, when the first live pictures came down, we gasped. Many of those troublesome heat shield tiles had come off. Referring to maneuvering rocket engine pods at the rear, "Looks like someone took big bites out of them," Young deadpanned.

The big gaps could be seen on TV. What could not be seen was the underside of the shuttle. If those tiles were damaged the ship would burn up on re-entry.

STS-1 Launch

After many hours, a NASA news conference led by the legendary flight director Gene Kranz of Apollo 13 fame ("failure is not an option") tried to assure us they were confident the underside was just fine. "How do you know?" I asked. "You can't see it."

"A combination of assets and assessments was brought to bear and it is our best judgment at this point in time that the mission should proceed and we expect it to come to a successful end, next question ..." Kranz replied. He would say no more.

Later, we discerned that various American intelligence devices, especially telescopes in the mountains of Hawaii had allowed someone to look Columbia over. But, we did not know that at the time, and every news story included collective worry that Young and Crippen might not make it home alive. Chris Kraft was watching it all and was worried. Had they overestimated what they could do?

Vice President George H.W. Bush called the crew in space with congratulations. President Ronald Reagan had been scheduled to do so but was recovering from an assassination attempt two weeks earlier.

A million miles and 37 orbits later, the first space shuttle mission ended in a safe landing on the dry lake bed at Edwards Air Force Base. In the media center, all pretense of detached neutrality was put aside. Bottles of champagne appeared and we all toasted America's return to space and cheered when Young and Crippen emerged triumphant.

Little did we know how badly some things had gone.

Some of the missing tiles did turn out to be in a critical area. A few more seconds of extreme heat and one of the landing gear doors might not have opened, leading to a crash landing. As it was, the door was badly warped by heat. Young later also said the landing gear itself was bent.

Those missing tiles had been knocked off by the explosive force of the ignition of the huge shuttle main engines and the mammoth force of the start of the solid rocket boosters. It caused what engineers term an overpressure. Most of us call that an explosion.

Whatever you call it, it also caused the main control body flap to bend so far that, had those details been known, it would have made them assume a safe landing was not possible.

Young was quoted later saying, "If we'd known about that I'd have flown to a safe altitude and we would eject." If that had happened, Columbia would have ended its mission in a crash.

But, it was a test flight. It ended with everyone safe and sound. It was hailed as the start of a new era in human spaceflight, and it was.

The celebrations went on for more than a month and included an international tour by Young and Crippen, the hero astronauts representing the best of America.

But, the problems on the first mission of Columbia would occasionally repeat themselves and be compounded by others. Eventually, they would even claim the great ship and her astronauts.

STS-2: Enthusiasm,
Big Improvements, Omens

The second flight of the space shuttle did not happen for seven months.

There were renewed efforts to fix the shuttle insulating tiles and many were removed and reattached. The damage caused by heat and the explosive force of engine ignition was repaired. They also changed launch procedures to dampen those same forces and not a single tile came off during the flight.

Planned to last longer, it was otherwise very similar to the first flight. It was the last flight to have a fuel tank painted white. Leaving the tank unpainted removed 600 pounds of weight. It has been estimated that it costs NASA more than $10,000 a pound to send something to orbit aboard a shuttle, so, the change to an orange tank saved more than $6 million a flight.

It also carried some payloads.

Various intelligence-related instruments including an imaging radar that can see items below the surface of Earth were downplayed.

It was the first flight of the robot arm. Called Canadarm after the nation that built it, testing was to be a highlight of the planned five-day mission. But, fuel cells that create critical power and drinking water began to fail and the mission was cut short.

Engle and Truly in Simulator

Still, the mood in mission control and throughout Johnson Space Center was joyous. The time between missions had allowed President Reagan to fully recover from the assassination attempt and he visited personally. He took a microphone in the room and congratulated Engle and Truly for making America proud.

Also in the room was veteran NASA photographer Andrew "Pat" Patnesky. He had captured images of history since Mercury days, but the joy of the moment and the personality of Reagan had his total attention. Perhaps that's why he stepped backwards and his thin frame folded neatly into a trash can right next to the commander in chief.

"There I was with my butt stuck in a trash can, real embarrassed but still shooting a picture. The president says, 'Are you OK?' and, when he realized I was, he just kept going so I could take my pictures. Talk about embarrassing ..."

Up in space and knowing they were to come home early in a few hours, commander Joe Engle and pilot Dick Truly spent part of their sleep period mostly awake testing that robot arm. They did that during periods when ground controllers were unable to monitor them due to a loss of signal and communications.

That was typical of Engle. Now a retired major general, he is a man of boundless energy, a top test pilot who earned Air Force astronaut wings flying the X-15 and had remained a NASA astronaut even when his scheduled seat to the moon was taken away. He likes to say, "90 percent of anything is enthusiasm."

But after launch the solid rocket boosters were inspected, and they found the first case of what was called "O-ring blowby" in a rocket joint. "The rocket is assembled in sections, and a series of rubber O-rings prevents hot gases from leaking through the joint," explains former NASA chief engineer Milton Silveira. Being the first time it leaked, they did some tests and made notes and said they would look for other instances, but would not change the design. During the next five years, O-ring blowby would happen several more times with no redesign until it destroyed Challenger and killed her crew.

For Engle and Truly, if 90 percent of their success was enthusiasm, the other 10 percent was a lot of hard work and luck.

Star Truck

STS-3: We Don't Put Our Pants On The Same Way

The third flight for the shuttle came nearly a year after the first. NASA was methodically reducing the time between missions as it had only one more flight scheduled before the spaceships were to be considered operational, not developmental or experimental.

With only minor issues that delayed the scheduled launch for an hour, this flight became the first to be extended in space. Commander Jack Lousma and pilot Gordon "Gordo" Fullerton were able to begin a series of previews of things to come with the shuttle.

They carried what was termed a getaway special payload in the payload bay. It allowed commercial and student-designed experiments, launches of small satellites and other uses with its GAS Can container. They carried several experiments on the shuttle middeck. Both men experienced space sickness, but were in orbit long enough for it to stop after a couple days. They also saw the $12 million shuttle toilet back up, something they did not talk about, and the first of a series of experimental medicine manufacturing, which got a lot of attention.

STS-3 Launch

STS-3 Lousma (left) and Fullerton

Columbia was scheduled to land on the dry lake bed at Edwards Air Force Base. Mother Nature did not cooperate. Late winter storms left the so-called dry lake bed flooded. The flight was extended to an eighth day to allow NASA to move support equipment from California to the still-dry White Sands Proving Grounds in New Mexico.

In space, the crew had put the shuttle and its robot arm through their paces. The tests continued all the way home. As Lousma said, "It got interesting."

He flew most of the return from orbit manually, but engaged the auto-land system, an automatic pilot much like that used on the Boeing 747, to give it a test. It had some problems and sent the shuttle toward its landing spot at too fast a speed, nearly 300 miles per hour. Lousma waited as long as possible to take over from the computers and that led to a spectacular landing that had controllers gasping in Houston.

Columbia was flying at 285 knots and only 150 feet above the ground when the landing gear finally appeared. Because of the high speed, the nose gear did not touch down for much of the runway, and efforts to bring it down failed with the shuttle nose bobbing back into the air despite Lousma's best efforts. First criticized by some for a less than picture-perfect touchdown, Lousma gained praise for his cool handling that allowed the engineers to better understand what they needed to improve in that autopilot system. It also helped convince every commander for the next 30 years landings would be flown by hand.

In their post-flight news conference, Gordo Fullerton showed some film they shot on the middeck. In weightlessness, he was seen taking a pair of pants in hand, lifting both legs and putting them on BOTH legs at the same time.

"See?" he deadpanned, "We astronauts do not put *our* pants on one leg at a time." The room erupted in laughter that captured the happiness that coursed through the space program at the end of the successful flight. Of course, being forced to land in the desert meant the shuttle main engines were badly damaged by blowing dust and would have to be replaced.

The shuttle was being touted as nearly "operational" and the key to making spaceflight safe and accessible for nearly anyone. These were heady times.

STS-4: The First Payback

While Armstrong and Aldrin were walking on the moon in 1969, the group planning the space shuttle had to make some significant changes in the shuttle design to accommodate an organization so secret that its very name was classified. It was dedicating a significant budget to underwrite the space shuttle and the National Reconnaissance Office, usually referred to as the NRO, joined the Air Force in assuming that all of its spy agency payloads would be launched from the manned spaceship.

Astronauts Ken Mattingly and Hank Hartsfield were flying not just the final shuttle developmental flight. It was the first partially-classified manned space mission in American history.

The public parts of the mission were nearly flawless.

For the first time, Columbia launched exactly on time and on schedule. It carried a series of student-designed small payloads, lasted the planned seven days and was the first mission to end on a concrete runway at Edwards Air Force Base. Surrounded by waving flags and patriotic symbols, President Ronald Reagan brought his wife, Nancy, along to welcome the crew home on July 4, 1982.

Out of public view, the flight tested an Air Force/NRO payload dubbed P82-1 that was a pair of devices testing the

STS-4 Hartsfield (left) and Mattingly

ability to detect missile tests from space. It failed due to a lens cover that would not open properly. The secrecy would increase during several NRO/military missions and keep from public view some of the most spectacular firsts for America in space.

But, as of that July Fourth in 1982, the development of the shuttle was pronounced over and its designation as an "operational spaceship" was declared.

The Space Truck

NASA was so anxious to prove the shuttle was truly operational that it was dubbed the Space Truck by managers and a few astronauts.

For the fifth mission, the ejection seats in Columbia were disabled and the crew wore jumpsuits instead of pressure suits. On STS-5, a pair of communications satellites was launched. It also carried the first four-man crew. Apollo-Soyuz astronaut Vance Brand was at the controls with pilot Bob Overmyer. With them were Joe Allen and Bill Lenoir. They were not pilot astronauts but mission specialists. The first planned shuttle spacewalk was canceled due to malfunctions in their spacesuits, but the mission was otherwise labeled a success. After landing, Columbia was sent back to the factory for major modifications.

The sixth shuttle mission marked the debut of the second space shuttle, Challenger. But, public interest was beginning to wane and the launch of a huge satellite and the first successful spacewalk were not headline-grabbers. The crew received the traditional presidential congratulatory phone call in space, but two million miles after liftoff, the landing at Edwards Air Force Base was worthy of but a few seconds of coverage on the evening news.

Columbia in Hangar

A Footnote in History

The seventh shuttle mission was a BIG deal publicly. It was the second flight of the newest shuttle, Challenger. But, more than that, it was the first time an American woman would make the trip to space.

"I hate it," is how Sally Ride told me she felt about all the attention to her gender and to her status as a first-ever.

Dr. Sally Ride is what one would expect. She is a brilliant and driven overachiever. She had risen to top positions in her educational and then professional life despite huge biases that existed about women in the workplace that were then common. She was a nationally ranked tennis player.

One of the first six women hired in 1978, the class chose as its group name TFNG that stands for Thirty-Five New Guys. From the beginning, these women did not wish to be considered different, and so Ride only grudgingly accepted what she privately described as "the girl stuff" to friends.

The preflight news conference was so well-attended that NASA had to move it to an auditorium that seated more than 1,000 people. It was a full house. Commander Bob Crippen described plans for the flight, how they would launch a pair of satellites, conduct experiments for European scientists and how medical doctor and mission specialist Norm Thagard would focus on space sickness.

Yeah, yeah, yeah and blah, blah, blah was the mood in the room and everyone knew why.

Sally Ride

The NASA public affairs officer called on me to ask the first question. Sally Ride and I had gotten to know each other and while I respected her drive to not be considered different, she respected my goal of telling a good story, and she was a good story.

"For Dr. Ride, as you know, someone has to be first and this morning it is me." The crew all laughed and it seemed to break the tension of Ride's anticipation.

"The crowd today shows just how much interest there is at the moment in you as the first American woman to fly in space. What's going through your mind as you look at this crowd today interested in you, someone who will one day be considered a footnote in history?"

Ride answered about her desire to do well on the mission and how she did not really see herself as different. She explained that the advent of the space shuttle meant women would take their place next to men in space.

The questions continued, most for Ride, but the answers stayed pretty much the same. A few weeks later they went to space.

The mission was a great success for NASA. The ship and crew performed well and the public was very interested. A few weeks later, before Ride and crew had what astronauts termed "my month in the barrel" of trips to several countries to do public relations, I received a piece of mail at my office. Inside was a handwritten note.

"John, thanks for your interest in the flight. It was great. Hope to see you soon." It was signed "Sally" and had a handwritten asterisk next to the name. At the bottom of the page, there was another asterisk. "A footnote in history."

STS-9: Rent-a-Shuttle

A highly modified Columbia was about to blaze another new trail in spaceflight. It had been worked over from nose to tail to make it capable of carrying Spacelab in its payload bay. This lab would house dozens of new experiments, and the shuttle would be able to stay in orbit longer than any previous shuttle mission.

For the first time in American history the crew would not be all Americans.

For the first time in history, the spacecraft would have active and awake astronauts working 24 hours a day.

And, for the first time in American history, the cost of the mission would not be paid by America.

What was dubbed the "rent-a-shuttle" carried the lab and conducted the science aboard mostly for the European Space Agency, ESA (pronounced EE-suh). Since the flight would be the most difficult and potentially risky since the first shuttle mission, NASA called on its ultimate astronaut to command one more time. It would be John Young's sixth and final mission to space.

As the astronauts gathered at Ellington Field in Houston to fly to Kennedy Space Center for the launch in their T-38s, Suzy Young came to see off her "Johnny," as only she could call him. With two big Afghan Hounds in the back seat of their car, she stayed with him as long as possible. Suzy knew that spaceflight was risky business and the simple fact was she did not want her husband to take that risk again. But, duty called, and John Young is a man with an unbending sense of loyalty and duty.

She waved as he taxied by and gave a good Navy salute. As he climbed into the sky at the controls of his jet, Suzy's eyes never left the speeding jet until it became first a small dot and then disappeared into the puffy clouds that hung over the Gulf of Mexico. She looked at me. "Thanks for being here," she said, and with tears welling, got in the car with her beloved dogs and drove away. She and other family members would fly separately to Florida in a week.

Columbia had experienced a few more firsts as she was prepared for the flight. She was the first to be rolled back to the vehicle assembly building and be taken off her stack of fuel tank and solid rocket boosters. One of the boosters had a defective rocket nozzle that was not discovered until it was on the launch pad. It was replaced, preventing what would have been a probable explosion.

At Kennedy Space Center, John Young and pilot Brewster Shaw were occasionally joined by four other crew members as they oversaw final preparation for flight and said their thank yous and shook the hands of as many technicians and engineers as possible. "I need to look them in the eye and want them to know who is going to be aboard that flying machine," he liked to say. For this veteran, spaceflight was an adventure, but the risk was known and personal and he wanted it to be that way with anyone associated with the mission.

Columbia rocketed to orbit in November 1983. Overall, the mission was a smashing success. The scientific work went very well. The new tracking and data relay satellite, or TDRSS, allowed for full-time communications, the transmission of high data rates as had never before been available to experiments in space. The payload specialists in the lab were connected to counterparts in a payload operations control center separate of mission control. There were first-time glitches, but they were minor. All went so well the mission was extended to 10 days, the longest flight to date.

Then, it was time to come home and those scary things Suzy Young worried about started happening.

Young and Shaw were at the controls four hours before planned landing. They fired one of the Columbia maneuvering thrusters to prepare for de-orbit and one of the main flight control computers crashed. A few minutes later, a second flight control computer crashed. Young immediately put the shuttle into free drift, meaning all of the controls were turned off while they scrambled to get the critical computers back online.

That second computer did finally reboot. Later, it was discovered that the shock of the thruster firing in space had jarred loose a piece of solder that shorted out the computers.

Of the incident, Young later said in his direct test pilot way of speaking, "Had we activated the backup flight control software, a complete loss of vehicle and crew would have resulted." With hobbled computers back to work, they headed for landing at Edwards Air Force Base. They had travelled more than 4 million miles during 166 orbits and were nearly home safe. Then, it got worse.

The flight control surfaces, the flaps and moving parts that control the shuttle as it flies through the air get their power from devices called auxiliary power units. The same devices are comnon on all commercial and most military jets. But, as Columbia neared the ground, two of the APUs caught fire and at least one

Tracking and Data Relay Satellite, or TDRSS

of them exploded in the engine compartment at the rear of the spaceship, just below the tail. Instruments showed problems, but no one knew what had really happened.

As Young and Shaw brought the Columbia to a halt on the runway, flames could be seen licking out of a vent just in front of the tail. The crew stayed aboard while ground crews checked. The fuel that powers those shuttle APUs is very dangerous and can be deadly to humans. It had leaked and was burning.

A NASA commentator in mission control issued a statement shortly that said the APU had exploded and was damaged and that it would be investigated. His bosses from NASA headquarters in Washington went ballistic.

This was not the kind of thing they wanted to hear about their "operational" fleet of spaceships. They issued a "corrected" statement that a "turbopump experienced a temporary overspeed resulting in some damage" and otherwise minimized the incident.

Astronauts and those of us who were more knowledgeable just laughed it off as the silliness from a PR guy. When the astronauts flew home to Houston and a big welcome party, Suzy Young ran to her hero husband and gave him a big hug. The tears this time were tears of joy. He would not take that risk again. It was not because he did not want to fly again. It would be a decision made after the final flight of Challenger.

We all celebrated the successful mission and simply felt vindicated in an "I told you so" sort of way when it was announced a week later that the APUs had indeed exploded and there was a lot of damage. Yeah, we know that and did not fall for the PR guy.

Little did we know just how deep that practice of rationalizing the risks and near-misses would become in the space agency. Little did we realize just how serious it could be to not squarely face the dangers discovered along the way and fix them. It would be a lesson learned the hard way. It would be taught by the first teacher in space.

Star Truck

The Dream Job

The early flights of the shuttle had been media extravaganzas for NASA. But, with the landing of Columbia closing out the ninth successful flight, the American networks rapidly lost interest. For a couple years they had competed with each other to be THE choice for viewers interested in space exploration. They brought in large crews and built studio facilities on borrowed land at Johnson and Kennedy Space Centers.

I worked for KHOU-TV, the CBS affiliate in Houston. I was the space reporter. This grand adventure was now my beat. With the withdrawal of much of the network coverage, I also became the eyes and ears on space for the network.

There would still be a network presence as they called it. Correspondents such as Bruce Hall and David Dow would fly in. I was jealous, of course, because they attended the launches and then rushed to Houston. I did not. As soon as the NASA mission commentator uttered the words, "the shuttle has cleared the tower …" at liftoff, the control and destiny of each flight was taking place in Houston. So, I stayed there.

The opportunity to get to know the various astronauts came fast and furious. It is an old joke that NASA's ability to take something so interesting, dangerous, expensive and where, if anything goes wrong people can die, and somehow make it seem boring shows just how good they are. Sometimes the descriptions, the deliberate management and engineering minds in charge were boring. But, the astronauts? They were some very remarkable people. The average astronaut is like the most driven overachiever most people will ever meet. I was paid to hang out with them.

John Getter and Jimmy Wong in Shuttle Mock-up

The next several missions were very interesting and showed just how versatile the space shuttle would be. They were the proof of Kraft's Law as NASA sometimes overestimated what engineers could accomplish in the short-run and underestimated what they could do in the long-run.

First, to help make it easy for paper pushers and schedulers, NASA changed the way it designated flights. Instead of STS-1 through STS-9, they began naming them after the makeup of their payload. So, STS-10 was canceled due to payload delays. The next mission was dubbed STS-41-B. It was a great example of making things a little more boring. The flight was exciting. The name? Not so much.

NASA had given me unprecedented access to the crew as they trained for the flight. They would deploy satellites, but the real excitement would come during a pair of spacewalks that would use what was called the manned maneuvering unit, the MMU. It was a jet-powered backpack that allowed an astronaut to fly free and unattached to the shuttle.

"Welcome to space. Let me show you how this will work," is how Bruce McCandless and Bob Stewart greeted me through underwater speakers in the big pool where astronauts train for spacewalks, or extravehicular activity, EVA. I was in scuba gear. They wore space suits. Being in the water simulated being in space, but it did not eliminate gravity.

Their suits still weighed 600 pounds, but they spent hundreds of hours practicing how to use tools, how to get in and out of the ship, how to put on the MMU and anything else that might arise. Of course, they also worked out emergency procedures should they need to repair the shuttle itself. But, they could not practice actually flying the MMU in the water.

It was amazing to walk into a room in Colorado and fly in space. Engineers had built a full-motion simulator for the MMU and I was invited to watch McCandless and Stewart practice. It was not easy. The MMU was flown along a curving curtain strung with small target points. The goal was to stay within an inch or so, but not touch them. We taped the astronauts getting their final workout, did an interview and McCandless ended it saying, "So, you wanna fly it?"

"Hell yes, I wanna fly it!"

The MMU was controlled by a pair of joysticks, one for each hand. It used nitrogen gas jets to move in any direction, up, down, forward, back and it could do front or back flips. But, if an astronaut ran out of gas before returning to the shuttle, he was doomed. They let me try a few simple tests and then challenged me to fly the curtain.

McCandless using MMU for EVA

The curtain represented what was called a sinusoidal wave. Dictionary.com defines that as "having a magnitude that varies as the sine of an independent variable." What that meant was that I had to fly a wave pattern that went up and down and simultaneously in and out.

I figured out that the MMU handled a lot like a wheelbarrow. For those not acquainted with that type of manual labor, a wheelbarrow has handles at the rear and a single wheel in the front. To make a turn, simply point the wheelbarrow where you wish to move your load of bricks or whatever. So, to move the device to the right, first rotate yourself toward the left so that the front is pointed right. Got it?

A few weeks later we watched the amazing sight of McCandless and Stewart doing the same thing. But this time, the curtain was a limitless black sky with a lone human being flying free. They were human spaceships. It produced one of the most iconic images in spaceflight.

The flight was the first to land on the runway at Kennedy Space Center. It ran right into a small flock of birds who apparently did not hear the big glider coming. Several weeks later a large envelope arrived. Inside was a picture of the landing. Commander Vance Brand included a mission patch flown in space and an official FAA Bird Strike/Incident Report form. In a handwritten entry it says:

Type of Aircraft – shuttle
Location – Shuttle Landing Facility, Titusville, Florida
Speed at Impact – 290 kts
Damage to aircraft – damage to tile on upper nose section
of shuttle Challenger
Evasive action taken – none

It is signed by "operator" Vance Brand.

FAA Bird Strike/Incident Report

The MMU flew two more times and both were spectacular.

On the next mission, STS-41-C, astronauts "Pinky" Nelson and "Ox" Van Hoften tried to grab a broken satellite called Solar Max out of orbit. But, the mechanical extension on the MMU designed to latch on – a maneuver I had been privileged to try in that Colorado simulator – did not work. The satellite was nearly lost, but retrieved by fellow crew member TJ Hart using the shuttle's robot arm. The satellite was repaired and relaunched. It was the first time a satellite had been caught and repaired in orbit – an ability that would soon prove nearly invaluable.

The mission was commanded by STS-1 veteran Bob Crippen. Challenger's pilot was Dick Scobee. He would later command the great ship on her final mission. But this flight was nothing but a spectacular success. It was also a hit movie. The flight became the basis for the IMAX spectacular "The Dream is Alive" that is still shown in theaters.

The MMU flew one more time. In the fall of 1984 it was aboard STS-51A and was used by astronauts Joe Allen and Dale Gardner to retrieve a pair of communications satellites whose failed rocket boosters had left them in useless orbits. Insurance company Booze Allen had insured the original launch and paid a significant part of the cost of the mission. The satellites were brought back to Earth, repaired and then launched aboard an unmanned rocket.

Tours, Toilets and Technology

As the space reporter at Johnson Space Center, I was called on to give some tours of our facilities and to take guests inside the areas where astronauts trained. We were often able to shoot video inside the actual mock-ups and that was also a favorite visitor spot. No one wanted to miss the chance to sit in the commander's seat of the space shuttle. And everyone wanted to know about the toilet. It is not easy to handle human body functions in the weightlessness of space. Astronaut Rusty Schweickart wrote a detailed article of the early days of defecation and urination in orbit for the Space Colonies edition of Whole Earth Catalog.

Years later, reminiscing with me on a tour of JSC, Chris Kraft had to chuckle as he recalled that the astronauts on a long Gemini two-man flight in a very cramped capsule had to shed all inhibitions. Relief required the removal of a pressure suit, floating out of the couch about two feet, placing a plastic bag strategically over one's "exhaust" and defecating, all within about 18 inches of the crewmate's face. "Spaceflight was always exciting, but not necessarily glamorous," he laughed.

The shuttle has a toilet. Developed at a cost of more than $20 million, it uses air flow to move both urine and feces to where they are supposed to be. It did not always work properly. Several toilet failures were reported and sometimes resulted in

Shuttle Bathroom

the gross and potentially dangerous problem of floating waste. During the years, it was pretty much perfected and generally did what it was supposed to do.

But, for those touring the mock-ups, the toilet was a highlight. Few of us will go to space, but all of us go to the bathroom and understanding how that happens is fascinating to all.

During the next two years, we saw the first flights of Discovery and Atlantis. There were classified missions for the

National Reconnaissance Office and the Pentagon. Satellites were launched. New science was explored. Missions got longer and seemingly more routine.

And, as everyone became more comfortable with the "routine" of space flight, a terrible thing began to reoccur that would only be recognized in hindsight. Twice.

In the early days of the manned space program, the successes of first Mercury and then Gemini missions led to hubris. It is a particularly terrible type of technological arrogance that can appear when engineers or others live out the first part of Kraft's Law. When those in the business of spaceflight overestimate what they can do in the short-term, people die.

The Apollo 1 fire occurred in no small part because overconfident mission planners and engineers counted on repeating the pattern of success. They hurried in the construction of the capsule and had astronauts working in a pure oxygen atmosphere inside that capsule. Any junior high science student can tell you that nearly everything can burn in pure oxygen.

On Jan. 27, 1967, astronauts Gus Grissom, Ed White and Roger Chaffee died on the launch pad when a spark in that pure oxygen caused a roaring burst of flame that consumed everything inside in less than 20 seconds.

Now, a new generation of spaceship was flying successfully. Sure, there were some problems and some close calls. However, no one wanted to hear of serious issues or worries.

No one wanted to say anything that might jeopardize funding. The mindset was "mission success" and that was what would be expected.

It would be one day short of the 19th anniversary of the Apollo 1 tragedy when all the world would be reminded what happens when space explorers overestimate what they can do.

Abort!

It was a typically hot summer day in July when Challenger was on the launch pad for her eighth flight, the 19th flight of the shuttle program. This was the second complicated 24-hour operation carrying Spacelab and a crew of seven astronauts. Gordon "Gordo" Fullerton was the commander.

Although Fullerton had flown in space only once before as pilot of STS-3, he was considered one of the best pilots of large aircraft in the NASA ranks. For years he had flown the infamous "Vomit Comet" airplane, a KC-135, the military tanker version of the workhorse Boeing 707 passenger plane. It would fly in huge arcs that would produce zero gravity in 30-second bursts. He had been a top test pilot. His skill at handling a huge airplane the way most pilots could handle a fighter is why he had been one of the first to fly Enterprise and Columbia. That, combined with his quiet, calm command style, led to his choice to lead the second Spacelab mission like the first that had been entrusted to the iconic John Young. Like Young, he is not a classic "right stuff" high-profile jet jockey who fancies sports cars. Gordo drove a VW Vanagon. He was very good and he is very cautious.

On the first attempt to launch, Challenger main engines automatically shut down after ignition a mere four seconds from liftoff. Engine No. 2 had an equipment failure and all three of the mains were suddenly turned off even though they had begun creating those huge plumes so familiar to anyone who has seen a launch. Seventeen days later, the crew was back aboard and the launch went off after a short delay for another technical problem. In both cases, managers and engineers were convinced all was fixed.

STS-51-F Crew

"Go for main engine start, main engine start, 3-2-1, solid rocket ignition and liftoff! Liftoff of Challenger and Spacelab 2 and the shuttle has cleared the tower," sent the spaceship on her way. She was travelling at more than 100 miles an hour before she even cleared that launch tower.

"Challenger, Houston, you're go at throttle-up," from mission control. "Roger," Fullerton replied. He was already five miles away and six miles up barely more than a minute into the flight. The ship was moving at more than 1,000 miles per hour.

Two minutes into the flight, the solid rocket boosters separated as planned. At some 5,000 miles an hour they were climbing a half-mile every second.

A few seconds later, now moving at more than a mile every second, mission control said, "Challenger, Houston, press to ATO," a standard phrase meaning that if one of the three engines fails they will plan on dumping excess fuel and struggling to minimum orbit instead of a dangerous emergency landing on a runway in Europe. "Roger."

Just as all seemed fine, Gordo called "Houston, center engine fail."

"We copy, standby," was the instant response. Less than two seconds later, "Abort, ATO, Challenger abort ATO."

"Roger, abort ATO."

In mission control and aboard the ship things were extremely tense, and the voice transmissions strikingly calm.

"Main engine limits to enable, Gordo."

"OK."

Mission control commentator Brian Welch told the world, "The crew still has the ability to make it to Zaragoza (Spain) should they lose another engine; however we now have two stable."

We learned later they were anything but stable, at least on the control panels reporting what their instruments said. The flight controller responsible for the engines saw another indicating serious malfunctions developing. Within seconds, the instruments would automatically shut that engine down, too, forcing the crew to attempt to land on that runway in Spain.

Aborting a mission to an emergency landing was something that was in the "trick bag" available to any space shuttle launch. It was also one of the most-feared eventualities any commander contemplated. As veteran NASA mission controller and Johnson Space Center director Jerry Griffin once told me privately, "It's an interesting thing to do while you're waiting to die." Like critical care doctors, astronauts and controllers thrive on black humor to avoid stressing out.

So, in mission control, they had about three seconds to decide whether or not to allow the automatic safety systems to shut down another engine or to override and hope Challenger would get to space without an engine exploding and killing everyone aboard. Moving at more than 3 miles per second, 60 miles up and more than 500 miles from the launch pad, Gordo had been dumping fuel to reduce weight in hopes of continuing to orbit.

They were past the point where a safe emergency landing could be attempted when mission control said, "Challenger, Houston. Main engines to inhibit."

"OK, inhibit," Gordo replied calmly after a short pause. He knew what that meant. His gauges also showed an engine in danger of exploding. But, he completely trusted his team aboard the ship and on the ground.

Less than two minutes later, Challenger was moving at 5 miles per second and was in orbit. Aboard the ship they awaited word on whether they could stay there or would immediately come home.

They stayed.

Unknown to nearly everyone outside a close circle within NASA, there had been several close calls during launches. They all involved the solid rocket boosters. Sometimes they leaked hot gases between their joints. Sometimes the nozzles that steered the thrust were damaged. It was just not discussed and remained unknown because the "solids" did not have the same level of instrumentation monitoring them real time as did the liquid-fueled "mains" that kept us all on the edge of our seats for this launch.

Despite the scary start, this became one of the most successful missions in the shuttle program.

The ship and crew performed so well, the flight was extended an extra day to allow for completion of the science and engineering work that had been initially delayed by the ATO. And, the coverage of the flight quickly turned to one of the commercial payloads they carried.

Star Truck

Cola Wars and A Teacher in Space

Cola wars went to space.

Pepsi and Coke both paid to have the crew test special cans of their colas as possible refreshments for people in space. At the post-flight news conference there was little attention paid to the near-death experience and many questions asked the crew to comment on which was better, Pepsi or Coke.

Pepsi and Coke Cans from Flight on Display at Smithsonian

Finally, a somewhat exasperated Gordo said, "In space there is no gravity and these are carbonated drinks. You can't burp in zero gravity because the liquid does not stay in the bottom of your stomach. Truth is, carbonated drinks can make you feel sick in space."

In the Aug. 9, 1985, edition of the JSC newsletter "Space News Roundup," a front-page story headlined simply "Limits to Inhibit" detailed how booster systems officer Jenny Howard had only seconds to make a critical decision and had done the right thing very well. In the culture of mission control, that is considered perfection, and recognition of such performance is not often singled out but always expected by NASA managers. She was rightfully praised for heroic work, and NASA quickly focused on the next flight on the schedule as it considered replacing the sensors that had triggered the abort to orbit.

On Page 2 of the same edition of the newsletter was a headline "Bush Announces Teacher in Space" with a first paragraph that said, "Vice President Bush announced July 19 that Sharon Christa McAuliffe would be the teacher to go into space aboard the space shuttle next January."

Just a Teacher

Christa McAuliffe smiled, shook my hand and touched my shoulder. We had just finished our last on-camera interview before her flight to space. As she turned to walk to a waiting T-38 trainer jet where mission commander Dick Scobee would fly her to Florida for their launch, she laughed, "Now come on, don't shoot me walking away; you know how this thing makes my butt look."

Videographer Jimmy Wong and I shared that laugh from what would be our last conversation with this schoolteacher who had also become a friend.

She climbed a ladder into the jet and we did not show her butt. Under the blue flight suit astronauts wore the women also had to wrap themselves in an adult diaper. There are no toilets aboard two-seat T-38s and the flight took a couple hours. That puffy appliance had become a running joke between us. Other jets carried pilot Mike Smith, mission specialists Ron McNair, El Onizuka and Judy Resnik along with McAuliffe's fellow payload specialist, Greg Jarvis.

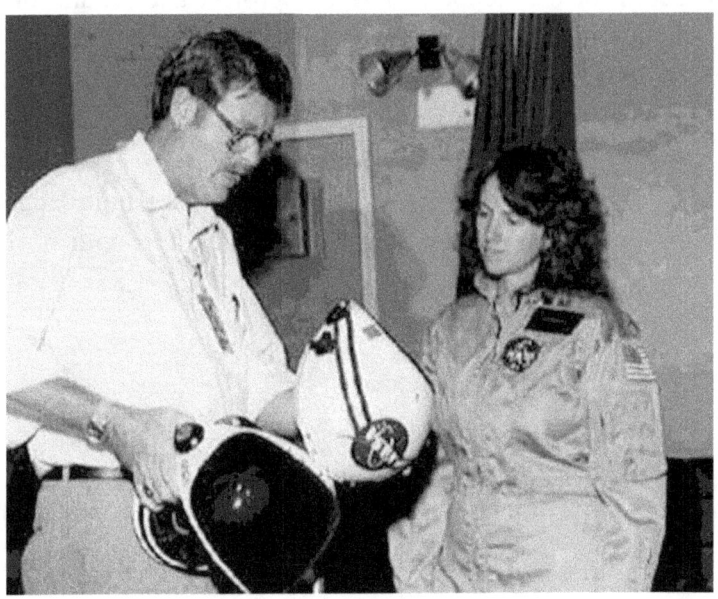

Christa McAuliffe Getting a Helmet Fit

I had first met Christa months earlier when she and the other final candidates for the Teacher in Space program had spent a week at Johnson Space Center as part of their selection process. All were bright and outgoing people and exceptional teachers. We shot video of their every move the first day they arrived. McAuliffe had just spoken with me and decided to duck into the restroom. "Now, you're not going to follow me in here are you?" she laughed. A running joke was forming.

During the next several months I was able to spend a lot of time with what was dubbed the 51-L crew. I had known Dick Scobee before, and he wrote a recommendation that I be considered for the Reporter in Space project that would follow this flight. Mike Smith was a top astronaut pilot with one of those smiles that lights up a room. I had seen El after work during other missions when we would talk over a cold light – always a light beer.

"That bottle is 45 minutes in the gym," he would laugh in recognition of a tendency to get a little chubby with no real effort. It was a trait we shared.

Back from left, El Onizuka, Christa McAuliffe, Greg Jarvis and Judy Resnik Front row from left, Mike Smith, Dick Scobee and Ron McNair

Judy Resnik and I had spent time together as I watched her train for her first shuttle mission. She had a wickedly sharp sense of humor. Once we were asked to act as judges for a gumbo cooking contest in the JSC community. Some of the samples were great.

Others? Not so much.

Her suggestion was to simply consider the BS factor of some samples. BS stood for baby shit, and she was right. That's how some of them looked and it was hard to taste them without laughing.

Judy Resnik

While in space the first time in 1984 aboard the maiden flight of Discovery, one of her main tasks was to test a huge solar panel system for the space station not yet built. It unfolded from a six-inch tall package to something approaching 100 feet. Working perfectly, it stood like a golden tower from the payload bay of the space shuttle.

Ron McNair was the Capcom speaking to the crew from mission control and inquired about the performance of the solar array. "Discovery, Houston, how's it look?" he asked in his straightforward style.

"It's up and it's big and it's stiffer than I thought!" she deadpanned. In mission control the double-entendre was lost on no one. The place filled with chuckles and McNair could be seen on TV putting his head down on his console to hide his face. She made people laugh a lot.

Jimmy and I had also had the chance to fly aboard the Vomit Comet several times with Ron McNair and Greg Jarvis. We flew to do news stories about whatever experiment or other work to be done on a future shuttle mission was being tested.

Greg had been bumped from two earlier shuttle missions to make room for politician-passengers. He was first replaced by Senator Jake Garn and then Congressman, later Senator Bill Nelson. His goal was to test the design of fuel tanks for satellites to assure the longest possible life spans for each. Without gravity, simply sucking all the liquid fuel out of a tank is complicated.

McNair flew because part of his duties for 51-L would be to capture the Teacher in Space mission on an IMAX camera for a feature film. Ron asked Jimmy for advice on how to shoot video and we considered it a great privilege to think some of what we knew would help them in space.

Ron McNair was a very earnest and brilliant physicist. Physics was not something at which many African-Americans had flourished at the time. He had spent his childhood picking tobacco and cotton by hand in the fields of South Carolina to help pay for college. Like Resnik, he was also a world-class musician. Judy was a classical pianist. Ron was a jazz saxophonist who had recorded with Jean Michelle Jarre.

For Jarre's upcoming album Rendez-Vous he was taking his sax to space and would record a solo that would be the first musical performance captured in orbit.

Ron McNair

The Tragedy

This flight was a public relations bonanza for NASA and a great news story for those of us who covered the shuttle missions.

The night before the planned launch, hundreds of friends and family gathered for a huge private party. Because launches were now considered so routine, my station and network would not be carrying it live. I was there as Dick Scobee's "bus captain" helping his wife, June, collect $20 from each guest to pay for the rented buses that would take us out to the private viewing area for the liftoff. It was a joyous evening. McAuliffe friends were sharing the same room and the place was full of optimism and excitement.

There were many glitches. For example, after the crew was inside the Challenger, the hatch was not sealing properly because a battery had gone dead in a portable electric screwdriver. Since all at NASA assumed few problems, there was no spare battery at the launch pad and it took many minutes to retrieve one and seal the hatch.

It was also gruesomely cold for a Florida morning. Huge icicles had formed alongside the spaceship because it was barely 19 degrees. It was very windy. But the skies were crystal clear and beautiful.

Out at our special viewing spot just a few miles from the launch pad, we struggled to keep warm and finally got off the bus just minutes before launch. I was very excited. This was the 25th mission I would cover, but it was the first time I would see a launch in person. Since it was the first time, my managers had decided no live coverage was needed, I was free to attend in Florida. To make things even better, I was a guest of the commander, and NASA had given me press badge No. 1.

As we watched the launch pad, I kept thinking of Christa's interview with me as she left Houston.

"I'm just a teacher and if I can do this, anyone can do this.

I never thought I'd be flying in space. But, you can't control what happens to you, just what you do about it. I hope people will see that if you just try, your dreams can come true, too."

Through the loudspeakers were heard, "3-2-1 and liftoff! Liftoff the 25th space shuttle mission and it has cleared the tower."

"Good roll program confirmed. Challenger now heading downrange," said commentator Steve Nesbitt from mission control.

I admit it. I was awestruck. What I had watched two dozen times on TV was nothing compared to the real thing. Despite the bright day, the intense power of the rocket engines was like looking directly into the sun. Because we were three miles away, there was no sound at liftoff. Then, seconds later, we could literally see the shockwave coming at us and the roar just kept increasing to deafening levels. My clothing vibrated because of the intense sound.

We stared at the sky barely able to hear the mission commentary.

"Twenty-two hundred 57 feet per second, altitude four-point-three nautical miles, downrange distance three nautical miles," Nesbitt reported.

Capcom Dick Covey called, "Challenger, go at throttle-up."

I heard my friend Dick say, "Roger, go at throttle-up."

Suddenly, from the ground we saw a large flash. It was confusing to even those of us who had watched many times. Was that the solid rocket booster separation? Had it already flown for two minutes? We could still hear the roar of the engines. Of course, sound travels much more slowly than light.

In a moment, all of the sound just stopped. What was heard on television was the sound of radios being destroyed. There was no real explosion. The ship shredded and fuel was suddenly ignited to a flash fire. There was no boom.

There was silence.

I was horrified as I realized there were pieces of spaceship flying everywhere. The solid rockets had flown on without a shuttle and were purposely exploded by safety officers. Pieces began splashing into the ocean.

"Holy shit," were the only words I could muster.

Behind us was the bus carrying Mike Smith's friends. A woman ran in circles screaming "Mikey, oh my God, Mikey!" as they were ordered back aboard their coach.

For several minutes we saw pieces tumbling. Some made huge splashes in the ocean. Parachutes carried rescue divers to the ocean from responding helicopters and some of the crew friends mistakenly thought the astronauts had bailed out. The crew had no parachutes and no way out.

We were ordered back aboard our bus by security officers. In the rear of the bus, one of Scobee's old friends was a retired Boeing test pilot with whom Dick had flown. "Well, Dick's going to have trouble explaining that he lost the aircraft," the old man laughed. He was in shock, as were we all and still did not realize Dick would be explaining nothing.

Some 50 people were sitting in absolute shock with disbelief on their faces trying to process what had just happened. As the "captain" for my friend, it was up to me to say something.

"Ladies and gentlemen, I think we all know what we have just seen. I suggest a moment of honor and prayer for our friends and their families who are suffering even more than we are. I hope you understand that the reason I am here is because

Dick and the crew told me they really respected my work. So, out of respect for them, I must now leave you and go to work."

Challenger Disintegrates

I put aside my launch guest credential to reveal press credential No. 1. The driver did not want to open the door without permission. I distracted him, grabbed the handle and walked off.

A stunned security officer ordered me back aboard the bus. Behind him, I saw a car race by driven by an ashen-faced John Young on his way to the families. "I need to go to work," I said.

Now he wondered if I should be arrested for being where journalists were not supposed to be. Finally, I convinced him to get me away from the friends and drop me at the press facility two miles away.

Less than five minutes after the launch, I walked into the CBS News building. Producer Margaret Erschler was seated beside correspondent Bruce Hall, and they were about to go with live coverage anchored by Dan Rather in New York.

"You OK?" she asked.

"Yeah. I need to be here."

"Sit down," she commanded and got up from her seat.

For the next three hours I produced the CBS coverage from Florida. My intimate knowledge of things paid off even that day. Seeing the replays of the launch, we were the first to report that images looked like something was wrong, that something was streaming from the shuttle before it disintegrated. Months later, the Rogers Commission that investigated the disaster would confirm fuel was leaking just as I recognized that day. As things began to calm slightly, I realized I really had to go to the bathroom and asked Margaret to take over. She did, but not without giving me a hug and saying, "I'm so sorry." She knew they were my friends.

Christa's words kept running through my head. "You can't control what happens to you, just what you do about it." She could not have meant her lesson this way, but this is how it was.

I went to the bathroom. I cried.

And, I worked every day for the next six months.

It's what my friends would have wanted.

Star Truck

The Aftermath

NASA initially responded with military precision to the disaster. Flight director Jay Greene ordered those in mission control to not touch anything. "Contingency operations" was his order. That meant nothing came in or left the room. Nothing was thrown away. Once you were told to leave, do so and don't touch anything. Leave it for the investigation. The investigation was devastating.

President Ronald Reagan led the nation in mourning. Johnson Space Center's beautiful campus was transformed overnight into an amphitheater. Thousands watched as the president spoke. The shocked families, the grieving friends, the stunned astronauts were surrounded by those who loved them and by an extended audience in the millions who watched from around the world.

After the service, the families of the crew retreated to Dick and June Scobee's house. Dick had been responsible for his crew and now June felt it her duty to be responsible for helping their families. With a cadre of astronauts, NASA security personnel and local law officers surrounding the place to protect their privacy, they gathered, grieved and wondered what to do now.

June later told me, "We really loved the flowers you sent." In what I felt was my feeble attempt to express personal pain I had asked the local florist to please just do something appropriate. They could not have done better.

"It was a beautiful collection of flowers, and in the middle was an American flag that was at half-staff. We were sitting around the living room table and decided we had to do something to continue the mission to reach our nation's children. We decided to call it Challenger Center and I reached down and raised that flag to the top of its pole," she told me. I am still stunned to have accidentally played such a role.

The president appointed former Secretary of State William Rogers to head the board that would investigate the loss. The Rogers Commission was made up of some of the best and brightest minds in America and they were relentless and unblinking in their look at how NASA itself was to blame for what had occurred.

First Came the Cover-ups

In an effort to soften the tragedy for the families of the crew, NASA managers told them the astronauts had died instantly in an explosion. Further, they said the bodies were destroyed and consumed by the fire and no remains would be found. Sadly, that was just not so and it was only under pressure from me and my CBS News colleagues the families were first told the truth.

We now know a solid rocket booster leaked at a seam. Hot gases and flames literally torched through a strut holding the booster to the huge fuel tank. When that strut broke, the booster tilted like a giant can opener and its nose punched a hole in the fuel tank.

Within microseconds the tank shredded, and leaked fuel and liquid oxygen were ignited by the shuttle main engines. The force of this skewed the shuttle sideways and it instantly disintegrated. But, the crew compartment, termed a pressure vessel, stayed intact. Inside, the astronauts continued to live until it slammed into the ocean a little more than two minutes later.

It took some time for the Navy ship salvaging the debris from the disaster to locate the crew compartment. We had developed tremendous sources, all of whom were worried that the truth would be covered up if they did not talk to us. My years of covering the program created enough trust they chose me to be the person to whom they leaked.

The astronauts undoubtedly died instantly upon colliding with the ocean. The forces of hitting the water at more than 200 miles an hour were tremendous and destructive. They shattered the compartment, but its miles of wiring kept most of it together as it sank in nearly 200 feet of water.

The USS Preserver located the crew cabin the day after it found a helmet floating on the surface. Divers discovered that the cabin was mostly intact and was filled with the remains of the crew. We knew this within hours although NASA continued to deny any of it.

It was the only time in my career that I held a story.

The next day we decided we were going to report the recovery of the cabin and bodies regardless of NASA denials. NASA public affairs officer Doug Ward only knew what he had been told, but he believed it. He and I argued passionately. I said I knew there were remains. He repeatedly said astronaut Bob Crippen said there were none and was understandably angry and frustrated. The JSC workers are very close knit and this was personal for him, as it was for me.

"Look," I finally said, "Today we are going to report on our newscast and then on Rather that bodies have been found.

I'll be goddamned if I'm going to let my friends hear about this on TV. Either someone go tell them or I'll drive over to the house and tell them myself!" I yelled.

"I'll call you back."

A few minutes later he did, and said astronaut Fred Gregory was going to Scobee's home. I told him I deeply appreciated it and that we would be proven right. He hung up. The families learned there were remains less than an hour before we reported it on television.

Aboard the ship off the Florida coast, the commander ordered his crew into dress uniforms. They would stand at attention and as an honor guard when the remains were brought back to shore. His order was overridden.

On directions from higher-ups, the remains were placed in nondescript containers some later described as trash cans. The ship waited offshore and returned with few running lights at night. The remains were loaded onto a vehicle and taken to a federal facility, out of reach of the local coroner who would later protest that such action broke Florida law.

The information being leaked to me continued to be shocking. There were a few more times I had to notify astronauts of what was coming so they could warn the families before it appeared on TV.

The astronauts did not die instantly. They were conscious as the crew compartment made a slow tumble to the surface. Someone, probably Judy Resnik, had pulled cables triggering emergency air supplies to the helmets of Scobee and Smith at the controls and her own as flight engineer. The air supply did not flow automatically. How do we know? It was an "on-demand" mask. Air is only provided when inhaled by a person wearing it. It had been breathed.

Confronted with the reality of this, cooler heads at NASA finally showed the crew full honors. Their remains were solemnly loaded aboard an aircraft in flag-draped coffins.

Meanwhile, the leaks just kept coming, often a day or two before the information was part of testimony to the Rogers Commission. We reported that evidence was a leaking seal had triggered the loss. Stories covered how previous near-misses had come close to destroying shuttles on earlier launches. For example, one year and two days earlier, Discovery was nearly lost due to an identical hot gas leak in a solid rocket booster launched on a bitter cold morning.

That information was kept secret from nearly everyone at Johnson Space Center and so was overlooked for this launch. Dedicated workers at Johnson Space Center were so angry they literally demanded I come to a dark room and view the video they discovered showing that near disaster and how it had been covered up in the name of "classified" information. I showed it on television that night, and the Rogers Commission took it up the next day.

I had learned about the infamous "O-ring" failure in cold weather the same way the commission did, only my lesson came over lunch at Frenchie's restaurant where I had earlier eaten with the crew themselves. An engineer had a piece of the O-ring. At room temperature it was flexible as it was supposed to be. Held in ice water in a glass for 30 seconds, it was stiff and prone to failure. Yes, he said. NASA knew this and had rules saying if it was cold the shuttle was not to launch. They waived the rule that fateful day because they did not know of the near-miss a year earlier.

The information presented to weeks of hearings was very high quality and shocking. There was the NASA manager who insisted, "We did everything right." Rogers looked at him incredulously and remarked that since a shuttle had been destroyed obviously something was not right.

How could it have happened? How could this group of world-class engineers, scientists and pilots have been involved in what turned out to be a torrent of cover-ups, denial and self-delusion? Perhaps the best explanation came from astronaut John Young.

"I've never known an engineer to die when his desk blew up," was his simple reply to that question. It summed up everything.

Wading through all the explanations and studies, it seems to boil down again to the first part of Kraft's Law: Engineers always overestimate what they can do in the short-run.

Carried to extremes, when that happens, people die.

Because of that, shuttles were grounded for more than two and a half years.

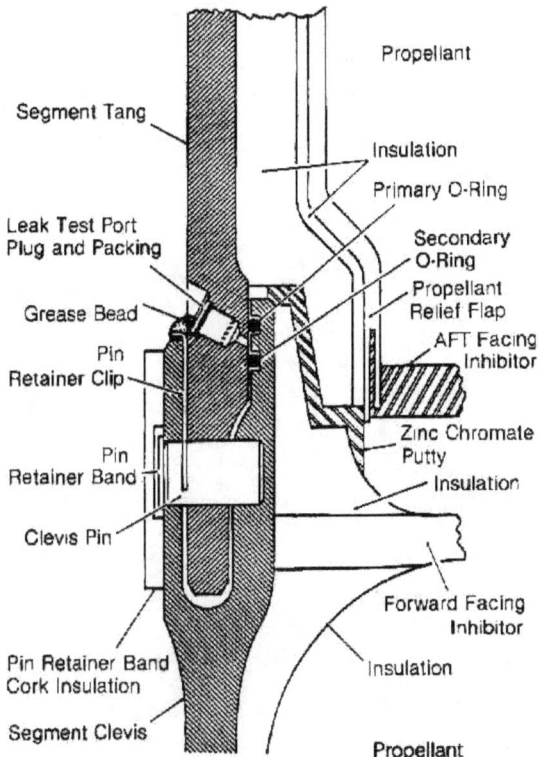

Figure 14
Solid Rocket Motor cross section shows positions of tang,
clevis and O-rings. Putty lines the joint on the side toward the
propellant.

O-Ring Diagram from Rogers Commission Report

A Funeral for Challenger

The wreckage of the spaceship had been handled like that
of any large airplane after a crash. The U.S. Navy did an

impressive job of recovering bits and pieces, large and small, and bringing them to shore. In an old hangar at Patrick Air Force Base, part of the Kennedy Space Center complex, they were carefully laid out on a grid that was painted on the floor. Everything was to scale.

Recovered Challenger Wreckage

For several months, engineers, scientists and others pored over the wreckage looking for any clues possible to find. And, they found a lot – everything except for one mysteriously missing wing. Reports to the Rogers Commission confirmed the great ship had essentially disintegrated in less than two seconds. It yawed sideways at such speed the skin was stripped off, the wings broke off – one of them apparently into countless tiny pieces, the tail fluttered to the ocean and the engine and crew compartments were shattered when they hit the water.

When they had learned all they could from the wreckage, NASA finally answered the demands of those of us covering the story and allowed us into the hangar. What we saw was breathtaking, literally and figuratively. The crew compartment was removed. They did not want to allow us to see where the astronauts had experienced approaching death between the time the ship went to pieces and their still-intact crew compartment smashed to the water.

But, everything else that could be found was there. The magnificent and beautiful ship we had known lay broken and bent, but so properly laid out as to be instantly recognizable. If that had not taken our breath away, the stench of the ocean did. Having been in the water for some time, the decaying remnants literally reeked.

For some, enough time had passed that they could look it all over, crack a joke and get the video or pictures they needed for a story. But, a few of us were solemn.

In front of us was a machine in which our friends had died. In front of us was the wreckage the creation of which had devastated families we knew and for whom we had great affection. And, Challenger – which had been such a symbol of great success, the daring audacity of humans willing to explore the unknown, which had thrilled us and scared us and made us so proud – was lying there mute and smashed.

After enough time had passed, NASA had to decide what to do with the wreckage. Many would expect it to be sent to a scrap yard. The aluminum and copper wiring could be reclaimed, the glass crushed and made into bottles and the rest would be just so much landfill. That simply would not do. Challenger was much more than just a broken pile of scrap. It was a broken dream that had seemed very alive to anyone who knew her. She would be buried.

With no public announcement, the day came to inter the remains. Through personal friends at NASA, I learned when it was scheduled and they agreed to allow me and my videographer, Jimmy Wong, to witness the event. It began shortly after sunrise.

There is an area of the base where minuteman missiles had once been kept in silos. There were test firings here of machines that could have signaled the end of civilization. Now, one of those decommissioned silos would be the tomb of a machine that represented some of the best intentions and greatest achievements of civilization.

The huge cement and steel door atop the silo was opened to a spectacular blue sky, much like that into which the ship had flown her last time.

One at a time, large white boxes bearing pieces of Challenger were lowered carefully deep into the ground. Workers who would normally do such heavy labor with boisterous energy as they shouted directions to one another were respectful and said little. They went about the business of burying an old friend.

When all the boxes were inside, the steel and concrete top was moved back into place and the few on hand gathered for a final moment of silence and personal prayers. They were saying farewell to a good friend.

Nearby, in another silo were stored the remains of the last fatal NASA spacecraft disaster. Apollo 1 had claimed three lives just a few miles away on a launch pad. Challenger took seven lives. They would never be forgotten and their burial was as solemn and heartfelt as for any American hero.

But, they did not claim the heart and soul of the space program. Just as had happened 19 years before, the wreckage of a beloved ship was surrounded by a growing commitment to make things right, to complete a mission, to make sure that these heroes had not died in vain.

We're Back!

Picking at lunch in the Building 3 cafeteria at Johnson Space Center, STS-26 commander Rick Hauck was speaking very softly. It was only a week after the loss of seven friends and colleagues. Next to him was his pilot, Dick Covey, who as Capcom that day had radioed, "Challenger, go at throttle-up!" and did not even realize there was a problem until many seconds later when a radar officer reported, "Tracking multiple targets," as the wreckage fell to Earth.

Hauck was still officially scheduled for flight within weeks. All knew it would not happen. They did not know much more. His mission was controversial and now weighed unbearably heavily on his mind. It involved launching a satellite using a liquid fueled rocket motor, meaning that the dangerous mix would be riding in the payload bay. Many astronauts did not like the idea at all and, with what had happened to Challenger, astronauts were beginning to reassert themselves.

At lunch, Hauck looked at me and said, "Only a fool would do something they know is more dangerous than it has to be, and I ain't no fool."

And his payload changed. NASA replaced the dangerous booster he had been slated to launch with a replacement for the critical communications satellite that was lost when the shuttle disaster occurred. Driven by the harsh lessons they had learned

STS-26 crew from left, Dave Hilmers, Dick Covey,
Pinky Nelson, Rick Hauck and Mike Lounge

in the aftermath, Hauck led a reconstituted crew made up entirely of veterans, the first time there were no rookies on an American spaceflight since Apollo 11. Hauck and Covey were joined by Dave Hilmers, Mike Lounge and Pinky Nelson. Together they traveled incessantly for two years even as they trained. They took to heart what John Young had done before STS-1.

Covey said, "I think we have met every individual and shaken the hand of anyone who is working on our launch. We want to make sure they know who is depending on them to do it right."

It took two and a half years for NASA to fly again. Discovery was back on the launch pad and the schedule was no longer being pushed as "operational" with planned liftoffs twice every month. Twenty-two missions were canceled.

Much had changed. NASA ditched the blue Nomex flight suits and astronauts again wore pressure suits. They installed a system that would allow, under some circumstances, for the crew to bail out during an emergency, a capability that had not been available since STS-5.

On Sept. 29, 1988, the countdown was under way and the ship lifted off to the cheers and tears of thousands on hand for the occasion.

"... 2-1 and liftoff! Liftoff! America's return to space as Discovery clears the tower," said Kennedy Space Center NASA commentator Hugh Harris.

It was his return to space commentary.

And, the same for Johnson Space Center commentator Steve Nesbitt. It was he who had the terrible job of announcing the loss of Challenger in January 1986. Those of us who knew both of them well had heard first some hesitancy and then joy from Harris, and it was the same for Nesbitt.

"Mark, one minute. Velocity 2,300 feet per second, altitude four-point-nine nautical miles, downrange distance three nautical miles." We could hear the tension that we all shared.

"Discovery, go at throttle-up."

"Roger go!" Hauck exclaimed.

Discovery given a "go" at throttle-up, three engines at 104 percent, velocity 3,200 feet per second, altitude 10.8 nautical miles, downrange distance eight nautical miles may read as a dry commentary. But there were very few dry eyes.

STS-26 Return to Flight Launch

Then, we were shocked at what we saw. Among the many things that were changed after the loss of the previous flight was a tremendous improvement in the quality of the tracking cameras. We could see details that had been hidden previously. And we saw flames surrounding the tail of the shuttle and its fuel tank.

"It's OK," NASA's Doug Ward said to me quietly. We were standing together in the grandstand next to the pad. I wanted to believe him.

Seconds later, the solid rocket boosters were released as scheduled and all was well. It turns out the "wake" of the shuttle and its tank actually draw some of the flaming main engine exhaust back toward the ship much like happens to the exhaust of a station wagon driving the highway with the rear window open.

Several minutes later, Discovery shut off its main engines and was in orbit.

"We're back," Ward smiled.

Getting Down to Work Up There

Chastened by the loss of a spaceship and crew, the NASA bureaucracy was collectively more safety focused even as they became more productive than ever. The astronauts had solidified their position as the most powerful group within the agency. Some professional managers and engineers hated that fact, but it was a fact.

There were several classified Department of Defense missions that were actually operating mainly for the National Reconnaissance Office. A pair of planetary probes was sent to Jupiter and Venus. They would unlock new secrets about our planetary neighbors. As one scientist later said of the liquid planet, "It is so dense that a cup of Jupiter weighs about as much as a small car on Earth." It was very interesting stuff and new science. Military and science work were the bread and butter of the shuttle program for a period, but military involvement was rapidly fading.

When it was first conceived, the shuttle was to be THE method of launching all satellites, civilian and military. When Challenger was lost, there were huge changes. Some spectacular capabilities such as the manned maneuvering unit, that jet backpack device, were deemed too dangerous to use. It was determined that it was not worth risking launching humans simply to launch machines like satellites that could ride atop unmanned rockets. The shuttle would be more specialized.

But the fact is, the high-profile and popular NASA space program was never cut out to be the stuff of secret spy missions and military work. For example, it is reported that during STS-27, a classified mission, the shuttle sustained serious damage to its heat shield tiles. Because of the classified flight, the crew could not send clear images of the damage to mission control for analysis that minimized the risks. The next time such damage would occur would be the last flight of Columbia in 2003.

In author Michael Cassutt's outstanding article "Secret Space Shuttles" for the Smithsonian's Air and Space magazine, he details the best information publicly available on those secret flights. (www.airspacemag.com/space-exploration/Secret-Space-Shuttles.html), For example, recounting STS-27, he says:

"The giant gold and silver satellite glittered against the black sky as space shuttle Atlantis closed in on it from below. Commander Hoot Gibson and pilot Guy Gardner flew the approach, while mission specialist Mike Mullane, at the other end of the flight deck, readied the shuttle's robot arm for a capture. Downstairs in the airlock, mission specialists Jerry Ross and Bill Shepherd waited in their spacesuits for Gibson's order to go outside and attempt a rescue.

"The mission of STS-27 had been to deploy the first in a series of new spy satellites that used radar to observe ground targets, in any kind of weather, day or night. But shortly after the astronauts released the spacecraft, called ONYX, from the shuttle's cargo bay on Dec. 2, 1988, one of its antenna dishes had failed to open. Without intervention by the crew, the billion-dollar satellite would become a hunk of space junk. As it turned out, they succeeded in grabbing, fixing and re-releasing

ONYX, for which they later received a medal from the U.S. intelligence community."

While the work by Ross and Shepherd was never publicly acknowledged, Ross played a key role in helping the astronauts who would later repair and refuel the Hubble space telescope. He explained to them just what it was like to do such repair work in a shuttle payload bay.

Earlier, before the loss of a shuttle, there had been what can only be termed a comical moment when the commander and crewmates for STS-10 renamed as 51-C visited a highly classified Air Force mission control center in Sunnyvale, Calif. Cassutt recounts:

(Commander Ken) Mattingly and three STS-51C crewmates – (El) Onizuka, Loren Shriver and Jim Buchli – had to take a trip to Sunnyvale. The astronauts were ordered to disguise their destination by filing a flight plan for Denver, then diverting to the San Francisco Bay area. They landed their T-38s at NASA's Ames Research Center in Mountain View, rented a "junky old car that could hardly run" according to Mattingly, and drove to an out-of-the-way motel arranged by their secretary. As they pulled up, Buchli, in the back seat, called a halt. "We made extra stops to make sure we wouldn't come here directly," he said. "We didn't tell our families, we didn't tell anybody where we are. Look at that motel." On the marquee was written "Welcome STS-51C Astronauts" with all four names in big type.

The classified mission was successful but had a terrible legacy. It was launched on the coldest day ever for a shuttle at that time. Its solid rocket boosters had serious "blowby" at a joint that nearly caused the loss of Discovery and her crew. Because it was top secret, however, none of that information was shared with most managers and astronauts at NASA, not even with the crew members themselves. One year and two days later, El Onizuka was again launched on an even colder day. He and seven others died when Challenger was destroyed by the same "blowby" that had nearly claimed him a year earlier. History proved that keeping secrets in manned spaceflight was not just difficult, it was sometimes deadly.

So, there were mixed emotions at JSC as DOD missions were coming to an end and satellites were being moved back to expendable unmanned rockets. More flights meant more opportunities to go to space. Less secrecy meant more safety.

Several missions were stunningly successful. There were those planetary probes, communications satellites and a lot of cutting-edge science accomplished. And, on the 35th mission came what history may judge the most important space shuttle mission ever. They deployed the Hubble space telescope. It would change the world.

Hubble Hobbled and Helped

Edwin Hubble was perhaps one of the most important scientists ever who was also unknown to most. Born in 1889, he was a classically eccentric genius. Born in a small Minnesota town, he was a champion high jumper, boxer and fly fisherman while still a teen. After a degree focused on astronomy, mathematics and philosophy at the University of Chicago, he became one of the first Rhodes Scholars and went to Oxford University in England. There, he studied jurisprudence before earning a masters degree in Spanish.

Returning to the United States, while affecting British mannerisms and tastes that would reportedly irritate students and colleagues the rest of his life, the Minnesota boy taught high school physics and math and was a basketball coach. He was admitted to the state bar in Kentucky, practicing law for less than a year. Bored, he completed his Doctorate of Astronomy at the University of Chicago. He served with honors in WWI and WWII achieving the rank of major. He received the Legion of Merit for work at the Aberdeen Proving Ground during the war.

Hired on as an astronomer at the then-new Mt. Wilson observatory, Hubble's research changed man's understanding of the cosmos. Until his work, it was thought the entire cosmos was contained in our galaxy, the Milky Way. Hubble proved that wrong by confirming the existence of other galaxies.

He also used what is called Doppler shift to prove the universe is expanding. That is called Hubble's Law.

Hubble died suddenly and unexpectedly at age 63 in 1953, at the very beginning of what would become known as the Space Age.

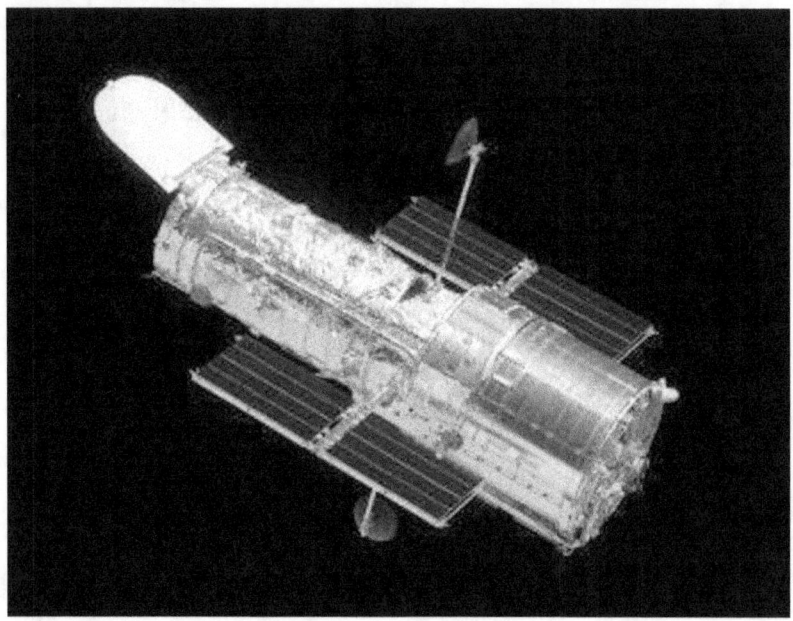

Hubble Space Telescope

In April 1990, Discovery carried what had been dubbed the Hubble space telescope into space. Astronaut Steve Hawley, an astronomer who considered Edwin Hubble a personal hero, had the honor of deploying what was called simply HST. Another member of the crew was pilot Charlie Bolden who would go on to become NASA administrator with the job of ending the space shuttle program.

The mission went well from a shuttle program perspective. The crew deployed the telescope. It passed some very early checkouts as they did other scientific work and safely returned home. But, weeks later news leaked out.

Hubble could not see clearly.

The huge lens made of thick glass was not perfect. It was distorting images. To make matters worse, it turned out that managers of the HST had known of the imperfections but did not raise red flags before launch. If Hubble was to be anything other than a tremendous disappointment and piece of space junk, it needed help.

Three and a half years and 28 shuttle flights later, help arrived aboard Endeavour. It was the most sophisticated and complicated shuttle mission in history. Two teams of spacewalking astronauts, Story Musgrave and Jeff Hoffman alternating with Kathy Thornton and Tom Akers, worked for hours in spacesuits, but only after commander Dick Covey and pilot Ken Bowersox had nudged up to the crippled satellite and captured it with the shuttle robot arm.

On the ground, controllers at Goddard Space Center in Maryland remotely stowed antennae and turned the satellite off. In orbit, it was obvious the solar arrays that provided electricity were damaged. They would be replaced along with the electronics they powered. The repairs were extensive, difficult, very time consuming and dangerous. And, they were completely successful.

The astronauts made five spacewalks during 11 days. They replaced the electronics of the telescope, the gyroscopes that could pinpoint places in space millions of light-years away, new compasses so the HST could navigate, better cameras to take advantage of the improved lens capabilities, It all worked.

The astronauts became some of the most famous in years. The spacewalking team of Musgrave, Hoffman, Thornton and Akers not only traveled the world speaking of the great achievement, they even appeared on the TV sitcom "Home Improvement" whose main character hosted a fictitious show dubbed "Tool Time."

It was another example of Kraft's Law. Engineers had overestimated what they could do in the short-term and the HST images were fuzzy and out of focus. But, they underestimated what they could do in the long-run. Critics said the repairs could not be done, were too dangerous and expensive.

They were wrong, and the astronauts accomplished more than they had initially believed. They accomplished so much, their work will continue to change man's understanding of our place in the universe after the shuttles are grounded forever.

Living and Dying in Space

There was a long and blurred flurry of shuttle flights that continued to do great science, engineering and exploration. Russian explorers and leaders had mourned the loss of the space station Mir as though it were an old friend. That is exactly what is had become to dozens of cosmonauts and many visitors, including American astronauts. It was an example of how to live and work in space, and it was outdated.

Launched in modules, the first segment of the complex was put into orbit by the Soviet Union in 1986. It was visited by astronauts from 12 countries even though it was not fully completed until 1996. Mir was occupied for 12 1/2 of its 15 years in space and had all the drama of a remote outpost. When the Soviet Union collapsed, cosmonaut Sergei Krikalev became known as the last citizen of the USSR as he spent more than 311 days in orbit awaiting a trip home courtesy of the then-new Russian space agency. Krikalev has spent more time in space than any other human: 803 days, nine hours and 39 minutes.

When the Mir was finally de-orbited in 2001, it had been worked very hard. There had been several visiting Americans and space shuttles. The first was the shuttle Atlantis that had first entered the fleet in 1985. That shuttle's first mission was, ironically, a classified mission for the Department of Defense to launch a satellite involved in monitoring the Soviet Union.

Mir

Mir had been crashed into by an errant space capsule, seen an onboard fire, lost major control functions and clearly demonstrated just how humans could adapt to bad developments and overcome them. Mir had been built by the Russian company RSC Energia, and its President Yuri Semyenov said it was the saddest day of his career when his engineers gave the command to crash it into the ocean. The same company had built literally every significant Soviet and Russian spacecraft beginning with Sputnik.

Meanwhile, Krikalev went on to be part of the first joint U.S./Russian shuttle flight and returned to orbit aboard a shuttle to help assemble the first segments of the International Space Station. He and astronaut Robert Cabana were the first people to enter the new station in orbit, literally turning on the lights and doing the initial checkout.

Every year, NASA would have another heart-wrenching Washington, D.C., battle to preserve funding. The shuttle was more expensive than had originally been planned, but it was, as astronaut John Young said simply, "A remarkable flying machine."

On its 28th mission, Columbia was lost, and once again the world was shocked as the deaths of its crew and the storied spaceship played out on television replays in February 2003. Unknown at the time, it was damaged by debris falling off its fuel tank during launch 16 days earlier.

On its first launch, Columbia had lost critical heat shield tiles and millions held their breath until it was safely home. This time, no one knew and few were watching the streak across the sky near Dallas.

As it re-entered the atmosphere and was flying just north of the astronauts' Texas homes, a hole in the wing burned through where a heat shield had been knocked off. Within milliseconds, the orbiter broke into pieces. Mercifully, as had not happened when a still slow-moving Challenger had disintegrated in 1986, the speed of the vehicle meant near-instant death for seven astronauts.

Once again, a special investigative group, dubbed the Columbia Accident Investigation Board, found the mechanical details of the failure, and as had happened with the earlier

High-Altitude Photo of Columbia Showing Disintegration

shuttle disaster, laid the blame at the feet of managers who had become complacent with the dangers of their spaceships.

Filed, Not Buried

In the 17 years between shuttle disasters, NASA had learned to handle the aftermath in a way that seemed to leave the public less stunned. Of course, the innocence of believing spaceflight was routine had been surrendered in 1986. In 2003, it was acknowledged that flying shuttles was dangerous stuff with inherent risks. The crew was honored on Earth and on Mars with a plaque aboard a robot explorer on the Red Planet. The wreckage of the Columbia, some 84,000 pieces, was not buried. The pieces were placed in an empty office complex at Kennedy Space Center.

Redesigns were undertaken. Managers' careers were ended. It would still take a two-and-a-half-year period to fix the shuttle system before they launched again, and again it was Discovery that restored American human access to space.

Star Truck

Killing the Dream

When they first flew, the shuttles were heralded as the follow-on to the great days of the Apollo program. They showed that America still led the world in technology and exploration. As a movie about them proclaimed, "The Dream is Alive!" Then a president killed it.

Space shuttles were designed and built to fly 100 times each. None of them would achieve that milestone. President Barack Obama confirmed he was grounding the shuttles after 135 flights by five spaceships. The once magnificent flying machines that survived would be parsed out to politically favored museums; the machines, simulators and trainers that had made history from Houston, Texas, for 30 years would be sent elsewhere and America would either rent seats from Russian Soyuz launches or from still-unproved and undemonstrated private sector spacecraft.

For the foreseeable future, Americans and America itself will be passengers in the final frontier, not leaders. The governments of Russia, India, France and China all have manned spacecraft being developed. The U.S. has unfocused plans for an Apollo-like capsule to take astronauts somewhere beyond Earth orbit someday.

No real goal. No real time frame. No dream.

What Was It Worth?

It took years to understand the real legacy and impact of the Apollo program. The most significant impact from that great space race challenge was the answer some engineers and scientists created to the question of how to control a small spacecraft landing on the moon a quarter million miles from home. That answer has since become the personal computer. It set off a wave of technology developments, data processing and storage capabilities that have forever changed the human experience.

It is a great example of Kraft's Law: they overestimated what they could do in the short-term from their engineering, but underestimated – in fact did not even predict – the effect they would have long-term simply because they needed a device to help control a capsule.

As this is written as the shuttle program ends, it is difficult to predict with certainty what the historians will say was the most significant product or impact of the glider named Enterprise and the spaceships dubbed Columbia, Challenger, Discovery, Atlantis and Endeavour. But, I have two strong possibilities.

First, the launch, repair and servicing of the Hubble space telescope not only dramatically showed how well people could work in space, it changed our view of ourselves. Already, scientists have discovered that when Hubble peers into space with a viewpoint roughly the size of a pencil, it can identify some 10 BILLION other galaxies. Remember, it was only in 1925 that Edwin Hubble first proved our Milky Way was not the only galaxy. In only a few years, his namesake in orbit has showed us there is an infinite number of galaxies.

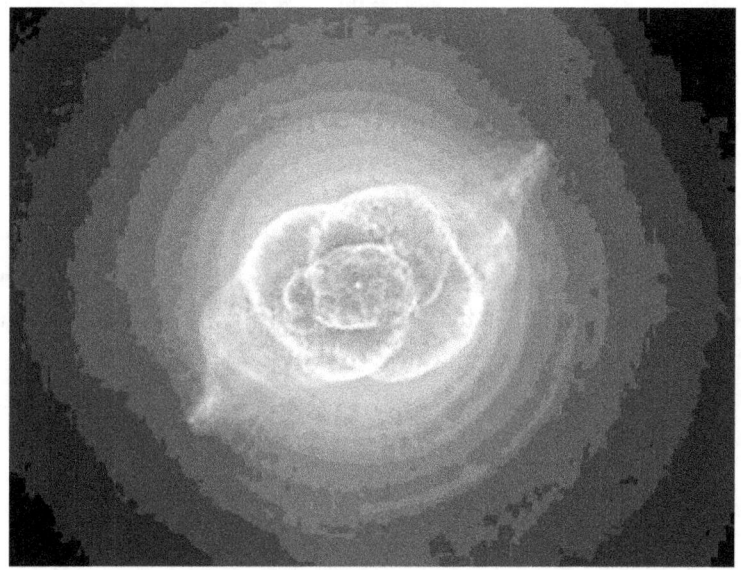

Hubble Picture of Cat's Eye Nebula
3,000 Light-Years from Earth

Further, it has shown there are other stars very similar to our sun, and that some are orbited by planets that could resemble our Earth. The Hubble telescope may not tell us if we are alone in the cosmos, but it has certainly humbled our sense of significance and uniqueness.

The PC and all the electronic capability, miniaturization, communication and freedom it has engendered has changed everything in ways we still struggle to understand. Thanks to Hubble, the laws of physics, our understanding of astronomy and human philosophy itself may be changed by what we are finding "out there" in ways we have yet to begin to understand.

Second, we may have quietly begun the epoch of human existence when people permanently live somewhere besides Earth. We will look back at the time of the shuttle as the time we used these magnificent flying machines to stabilize and make permanent our ability to live in low Earth orbit. From there we had a spectacular view of our own future: back to the moon, on to Mars, exploring our solar system, eventually placing permanent settlements on other planets, even looking beyond those outer reaches to see what is next. It all began with what were called space shuttles.

They were a glider and five spaceships that changed everything.

Epilogue

As one who grew up in what we have long called the Space Age, it is disheartening to see the American government effectively choose to stop leading exploration of space. I had the great privilege of covering NASA for many years, and also working as an executive for a small company called SPACEHAB (www.spacehab.com) now Astrotech that created great new capabilities for the space shuttle, but also saw its most creative proposals crushed by a government space bureaucracy. We brought together American and Russian operations in new ways, but proved to be too soon and too fast for government work.

We were not alone. Even a company called United Space Alliance, or USA, (www.unitedspacealliance.com) that was formed to carry out NASA shuttle operations saw its proposal to lease the shuttle Columbia dashed for various reasons. As one USA executive said at the time, "NASA knew we would show the reason spaceflight is so costly is the cost of the agency, not of the flights."

True or not, NASA was, ironically, more difficult for private enterprise to work with than the Russian space agency Roscosmos. It is they with whom I helped partner private-sector advertising and promotion in space and with whom a company called Space Adventures (www.spaceadventures.com) earned hundreds of millions of dollars flying the first space tourists to space stations.

So I am very pleased to see the development of several companies that plan to take paying customers to space, while other companies plan to build private facilities in which those passengers can work and play. But, I believe the decision to stop flying shuttles is premature and potentially grave for the United States.

Throughout history, the great nation of each time led the world in whatever exploration was taking place at the time. Exploration is basic to the human experience, and it is great explorers who achieve great things. It was true for the Greeks, Romans, Chinese, Portuguese, Spanish, English and the rest. And, it is equally true that when those civilizations stopped exploring, they declined.

NASA represents the American commitment to exploration and all that goes with it. As John Kennedy said of his reasons to go to the moon:

"We choose to go to the moon in this decade and do the other things, not because they are easy, but because they are hard, because that goal will serve to organize and measure the best of our energies and skills, because that challenge is one that we are unwilling to postpone, and one which we intend to win, and the others, too ... But if I were to say, my fellow citizens, that we shall send men to the moon 240,000 miles away ... and do all this, and do it right, and do it first before this decade is out – then we must be bold.

"Many years ago the great British explorer George Mallory, who was to die on Mount Everest, was asked why did he want to climb it. He said, 'Because it is there.'

"Well, space is there, and we're going to climb it, and the moon and the planets are there, and new hopes for knowledge

and peace are there. And, therefore, as we set sail we ask God's blessing on the most hazardous and dangerous and greatest adventure on which man has ever embarked."

Will America be bold? Will it face the greatest and most dangerous challenges to test itself as it leads mankind into the frontier of space, of exploration and human development? Or, will it choose to stay home, to be safe, to rent space and travel from others and NOT do the other things because they are hard, and we want to take it easy?

There was a time when we faced our fears and the unknown and looming disaster with the simple phrase, "Failure is not an option!" For the sake of America and the world itself, we must not just say it again.

We must live it.

Star Truck

To contact John for keynotes, inspirational talks, effective communications and consultations:

John Getter
6380 Golden Goose Lane
Las Vegas, NV 89118
702-531-7486
John@JohnGetter.com
www.JohnGetter.com

Star Truck

John Getter's Moonwalkers™ Series

Available for e-readers at:

Kindle and Kindle Apps: www.Amazon.com

Nook and Nook Apps: www.BarnesandNoble.com

All other e-readers and formats: www.Smashwords.com

Paperbacks available at: www.Amazon.com

Star Truck

For questions, custom covers, fundraising, or bulk purchases for schools, nonprofits and fundraising please contact:

Premiere Projects
Henderson, NV
24/7 GoogleVoice: 702-900-2176
www.PremiereProjects.com

www.ingramcontent.com/pod-product-compliance
Lightning Source LLC
Chambersburg PA
CBHW051536170526
45165CB00002B/763